MIX
Papier aus verantwortungsvollen Quellen
Paper from responsible sources
FSC® C105338

Humaira Khatoon
Jawad Nasir
Muhammad Shafiq
Syed Hussain Haider Rizvi

Nanotechnology

Synthesis techniques of silver nanoparticles

Anchor Academic
Publishing

Khatoon, Humaira, Nasir, Jawad, Shafiq, Muhammad, Rizvi, Syed Hussain Haider: Nanotechnology. Synthesis techniques of silver nanoparticles, Hamburg, Anchor Academic Publishing 2016

Buch-ISBN: 978-3-96067-088-9
PDF-eBook-ISBN: 978-3-96067-588-4
Druck/Herstellung: Anchor Academic Publishing, Hamburg, 2016

Bibliografische Information der Deutschen Nationalbibliothek:
Die Deutsche Nationalbibliothek verzeichnet diese Publikation in der Deutschen Nationalbibliografie; detaillierte bibliografische Daten sind im Internet über http://dnb.d-nb.de abrufbar.

Bibliographical Information of the German National Library:
The German National Library lists this publication in the German National Bibliography. Detailed bibliographic data can be found at: http://dnb.d-nb.de

All rights reserved. This publication may not be reproduced, stored in a retrieval system or transmitted, in any form or by any means, electronic, mechanical, photocopying, recording or otherwise, without the prior permission of the publishers.

Das Werk einschließlich aller seiner Teile ist urheberrechtlich geschützt. Jede Verwertung außerhalb der Grenzen des Urheberrechtsgesetzes ist ohne Zustimmung des Verlages unzulässig und strafbar. Dies gilt insbesondere für Vervielfältigungen, Übersetzungen, Mikroverfilmungen und die Einspeicherung und Bearbeitung in elektronischen Systemen.

Die Wiedergabe von Gebrauchsnamen, Handelsnamen, Warenbezeichnungen usw. in diesem Werk berechtigt auch ohne besondere Kennzeichnung nicht zu der Annahme, dass solche Namen im Sinne der Warenzeichen- und Markenschutz-Gesetzgebung als frei zu betrachten wären und daher von jedermann benutzt werden dürften.

Die Informationen in diesem Werk wurden mit Sorgfalt erarbeitet. Dennoch können Fehler nicht vollständig ausgeschlossen werden und die Diplomica Verlag GmbH, die Autoren oder Übersetzer übernehmen keine juristische Verantwortung oder irgendeine Haftung für evtl. verbliebene fehlerhafte Angaben und deren Folgen.

Alle Rechte vorbehalten

© Anchor Academic Publishing, Imprint der Diplomica Verlag GmbH
Hermannstal 119k, 22119 Hamburg
http://www.diplomica-verlag.de, Hamburg 2016
Printed in Germany

TABLE OF CONTENTS

TABLE OF CONTENTS i
LIST OF ABBREVIATIONS iii
PREFACE v
CHAPTER 1 INTRODUCTION 1
   1.1   NANOPARTICLES (NPs) 2
   1.2   TYPES OF NANOPARTICLES 4
   1.3   OPTICAL PROPERTY OF NANOPARTICLES 4
   1.4   SILVER NANOPARTICLES (AgNPs) 5
   1.5   METHODS OF NANOPARTICLE SYNTHESIS 6
      1.5.1   CHEMICAL REDUCTION METHOD 7
      1.5.2   BIO-BASED METHODS 8
   1.6   DETECTION OF SILVER NANOPARTICLES 10
      1.6.1   UV-VISIBLE SPECTROPHOTOMETRY 10
   1.7   SILVER NANOPARTICLES AND ANTIMICROBIAL ASSAY 11
CHAPTER 2 AIM OF THE RESEARCH 13
CHAPTER 3 MATERIALS AND METHODS 14
   3.1   MATERIALS 14
      3.1.1   REAGENTS 14
      3.1.2   SOURCE OF MICRO-ORGANISMS 14
      3.1.3   COMPOSITION OF SDB (Sabouraud Dextrose Broth) 14
      3.1.4   CHEMICAL COMPOSITION OF SDA (Sabouraud Dextrose Agar) 14
      3.1.5   CHEMICAL COMPOSITION OF NUTRIENT AGAR 14
   3.2   SYNTHESIS OF SILVER NANOPARTICLES (AgNPs) 15
      3.2.1   CHEMICAL REDUCTION METHOD (TURKEVICH METHOD) 15
      3.2.2   BIOLOGICAL METHOD (BY FUNGUS *TRICHODERMA PSEUDOKONINGII*) 15
   3.3   CHARACTERIZATION OF SILVER NANOPARTICLES 18
   3.4   ANTIMICROBIAL ASSAYS 18
      3.4.1   PREPARATION OF SDA PLATES FOR ANTIFUNGAL ASSAY 18

- 3.4.2 PREPARATION OF NUTRIENT AGAR PLATES FOR ANTIBACTERIAL ASSAY 19
- CHAPTER 4 RESULTS AND DISCUSSION 20
  - 4.1 SYNTHESIS OF AgNPs BY CHEMICAL REDUCTION METHOD 20
    - 4.1.1 OPTICAL SPECTROSCOPY MEASUREMENTS 20
  - 4.2 BIOSYNTHESIS OF SILVER NANOPARTICLES BY FUNGUS (*TRICHODERMA PSEUDOKONINGII*) 22
    - 4.2.1 OPTICAL SPECTROSCOPY MEASUREMENTS 22
  - 4.3 ANTIMICROBIAL ASSAY 23
    - 4.3.1 ANTIFUNGAL ASSAY OF SILVER NANOPARTICLES 24
    - 4.3.2 ANTIBBACTERIAL ASSAYS OF SILVER NANOPARTICLES 24
- CHAPTER 5 CONCLUSION 26
- BIBLIOGRAPHY 28
- APPENDIX 1 31
- APPENDIX 2 33
- APPENDIX 3 35

LIST OF ABBREVIATIONS

AgNPs	…………………....	Silver nanoparticles
$AgNO_3$	………………………..	Silver nitrate
Ag^+	………………………..	Silver ion
Ag^0	………………………..	Silver nanoparticle
°C	………………………..	Degree centigrade
C. albicans	………………………..	*Candida albicans*
DNA	………………………..	Deoxyribonucleic acid
E. coli	……………………………..	*Escherichia coli*
gm	………………………..	Gram
ml	………………………..	Milliliter
mM	………………………..	Milli Molar
mRNA	………………………..	Messenger Ribonucleic acid
NA	………………………..	Nutrient Agar
nm	………………………..	Nanometer
NPs	………………………..	Nanoparticles
rpm	………………………..	Revolutions per minute
S. aureus	………………………..	*Staphylococcus aureus*
SDA	………………………..	Sabouraud Dextrose Agar
Sol.	………………………..	Solution
SDB	………………………..	Sabouraud Dextrose Broth
T. pseudokoningii	……………………….	*Trichoderma pseudokoningii*
UV-Vis	………………………..	Ultraviolet-Visible

Dedicated to our families

PREFACE

In modern-day, Nanotechnology is being widely used in various domains of science. It deals with the nanoparticles having the size of 1-100 nm in one dimension and is used significantly in medical chemistry, atomic physics among other scientific disciplines. The synthesis of nanomaterial is of current interest due to their wide variety of applications in fields such as electronics, photonics, catalysis, medicine, etc. The applications of nanotechnology are growing owing to the fact that matter at the nanometer scale has different properties as compared with the bulk state. For this reason, many research groups around the world are trying new methods of synthesis of different materials at the nanoscale. Silver Nanoparticles (AgNPs) have been the matter of focus of researchers due to their unique properties (e.g. size and shape depending optical, antimicrobial, and electrical properties). A variety of preparation techniques have been reported for the synthesis of AgNPs. In this study, AgNPs have been chemically and biologically synthesized. For chemical synthesis of the AgNPs, silver nitrate is taken as the metal precursor and sodium citrate as reducing agent. Biologically, the silver nanoparticles are synthesized by using a fungus named as _Trichoderma pseudokoningii_. In biosynthesis of AgNPs the fungus is exposed to the silver nitrate solution. In this process the silver ions (Ag^+) are reduced to the metallic silver nanoparticles (AgNPs). The use of sodium citrate and biological entity make the processes non-toxic, cost effective and environmental friendly or green methods. All the samples are characterized by UV-visible spectroscopy. AgNPs exhibited a characteristic surface plasmon resonance band that is measured by UV-Vis spectroscopy, showing typical absorbance peaks for nanoparticles centered at 420 and 450 nm in different samples. The antimicrobial activities of silver nanoparticles are detected by the zone of inhibition against certain microorganisms by bore well method. The antifungal activities of synthesized nanoparticles are seen after 24 hours of incubation at 37 degrees Celsius of _Candida albicans_ growth on sabouraud dextrose agar and antibacterial activities of _Staphylococcus aureus_ and _Escherichia coli_ are seen on nutrient agar. As a result, _C. albicans_, _S. aureus_ and _E. coli_ are shown to be substantially inhibited by AgNPs. These results suggest that AgNPs can be used as an effective antimicrobial material. The microbicidal effects of silver nanoparticles are detected by the inhibition zone in bore well method. Microbial sensitivity to nanoparticles is a key factor suitable for long life application in a variety of scientific fields.

CHAPTER 1
INTRODUCTION

Nanotechnology is a significant field of modern scientific research dealing with design, synthesis, and manipulation of particle structures ranging from approximately 1-100 nm (Iravani, *et al*. 2014). It is a science that deals with matter at a scale of 1 billionth of a meter (i.e. 10^{-9}m = 1nm).It also deals with manipulating of matter at the atomic, molecular or supramolecular scale (Horikoshi, Serpone, 2013).

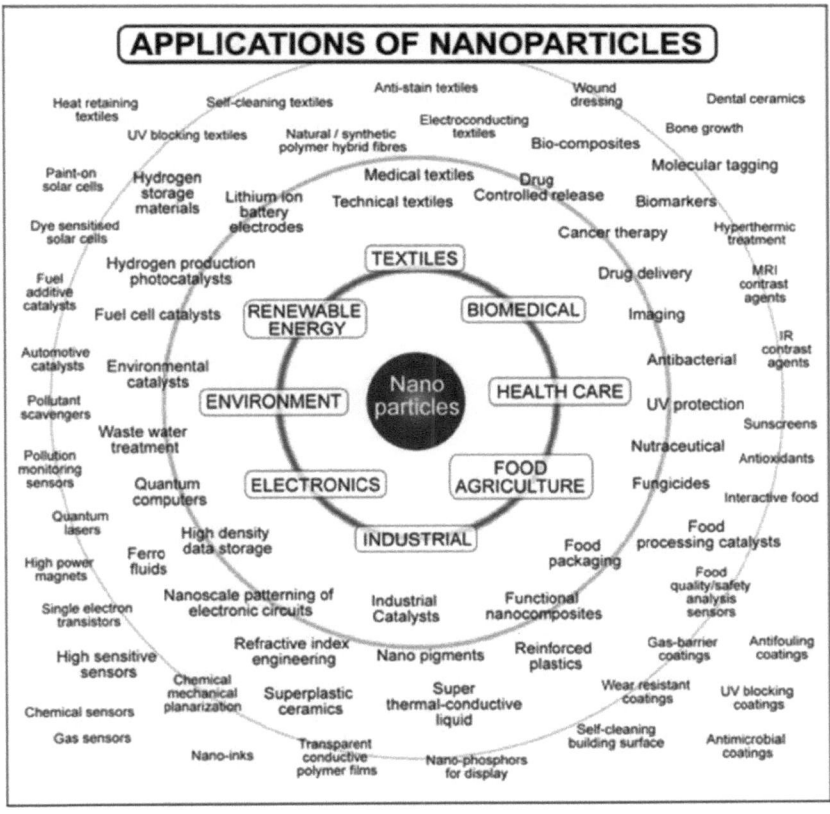

Figure 1.1: Applications of nanoparticles (Tsuzuki, T. 2009).

1

Nanotechnology is a multidisciplinary scientific area that concern with prevailing approach, facility and technology available in ordinary as well as modern routes of physics, engineering, chemistry and biology. For disclosing the inherent nanoscale activities of biological factors of living cells it was necessary to apply nanotechnology in life science, such applications are termed as nanobiotechnology.

The convergence of nanotechnology and biology or intersection between engineering and molecular biology or the application of nanotechnology in biological fields is known as *Nanobiotechnology* (Morais, *et al*. 2014; Fortina, *et al*. 2005). Nano biotechnology has turned up as a fundamental division of modern nanotechnology and due to its abundant applications and new approaches in the fields of material science receiving global attention.

Nanotechnology as well as nanobiotechnology is a fast rising scientific area of constructing and producing devices. An essential field of research in nanotechnology is the synthesis of nanoparticles (NPs) with different sizes, chemical compositions and morphologies, and controlled dispersities, therefore they are rapidly gaining importance in a vast variety of areas such as healthcare, food and feed, cosmetics, environmental health, optics, mechanics, biomedical sciences, chemical industries, space industries, electronics, energy science, drug gene delivery, optoelectronics, catalysis, reprography, light emitters, single electron transistor, non-linear optical devices and photo-electrochemical applications etc.as shown in figure 1.1 (Awwad, *et al*. 2013; Panigrahi, 2013; Iravani, *et al*. 2014).

1.1   NANOPARTICLES (NPs)

NANOPARTICLES (NPs) are particles which size spans the range between 1 and 100 nanometers, at least in one of the three possible dimensions. In nanotechnology, *a particle is defined as a small object that behaves as a whole unit with respect to its transport and properties* (Panigrahi, 2013).

In this size range, the physical, chemical and biological properties of the nanoparticles changes in essential ways from the properties of an atom or molecules and of the consistent bulk materials (Taylor, *et al*. 2013).

Nanoparticles are made up of materials of diverse chemical nature; the most common are metals, metal oxides, polymers, non-oxide ceramics, biomolecules, carbon and organics. There are several different morphologies of nanoparticles like spheres, cylinders, platelets and tubes etc. To meet the needs of specific applications, the nanoparticles are designed with surface modifications. Due to the vast variety of the nanoparticles arising from their wide chemical nature, morphologies and shapes, the medium of particles in which they are present, the state of dispersion of nanoparticles and most prominently, the various possible surface modifications, the nanoparticles can be subjected as an important and active field of science now-a-days (See figure 1.2) (Panigrahi, 2013).

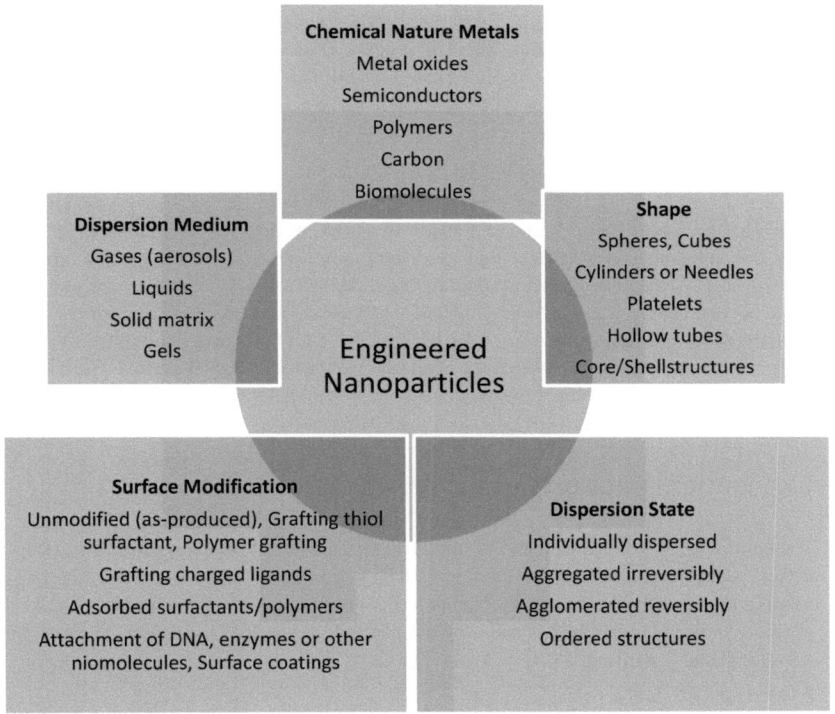

Figure 1.2: Various features contributing to the diversity of engineered nanoparticles. The same chemical can generate a wide variety of nanoparticles (modified from Nagarajan, *et al.* 2008).

Nanoparticles are of pronounced scientific concerns as they are, in effect, a bridge between bulk materials and atomic or molecular structures. A bulk material has constant physical properties regardless of its size, but at the Nano-scale, there are often observed size-dependent properties. Thus, the percentage of atoms at the surface of a material change as their size change at Nano-scale is important while bulk materials greater than one micrometer (or micron), the percentage of atoms at the surface is insignificant regarding number of atoms in bulk material (Taylor, *et al.* 2013). The conversion from micro particles to nanoparticles can lead to a variety of changes in physical properties. Two of the major factors are: the increase in the ratio of surface area to volume, and the size of the particle moving into the realm quantum effects predominate.

The increase in the surface-area-to-volume ratio, which is a gradual progression as the particle gets smaller, leads to an increasing dominance of the behavior of atoms on the surface of the particle over that of those in the interior of the particle. This affects both the properties of the particle in isolation and its interaction with other material (Landage, *et al.* 2014).

1.2   TYPES OF NANOPARTICLES

Nanoparticles can be grouped into two major categories, namely, *Organic Nanoparticles* e.g. carbon nanoparticles (fullerenes) and *Inorganic Nanoparticles*, which include magnetic nanoparticles, semiconductor nanoparticles, (like titanium oxide and zinc oxide) and noble metal nanoparticles *e.g.* gold (Au) and silver (Ag). Nowadays, noble metal nanoparticles (gold and silver) are providing exclusive material properties with functional versatility. Because of their size, type and advantages over available chemicals, these inorganic particles are recognized as important tools for medical imaging as well as for disease treatment (Panigrahi, 2013).

1.3   OPTICAL PROPERTY OF NANOPARTICLES

The optical property is one of the most important characteristic of nanoparticles. For example, a Gold (Au) nanoparticle of a 20nm sized has a specific wine red color. A Silver (Ag) nanoparticle is yellowish grey while Platinum (Pt) and Palladium (Pd) nanoparticles are black in color. These colors can be controlled by the size and the shape of a nanoparticle, see Figure 1.3 (Horikoshi, Serpone, 2013).

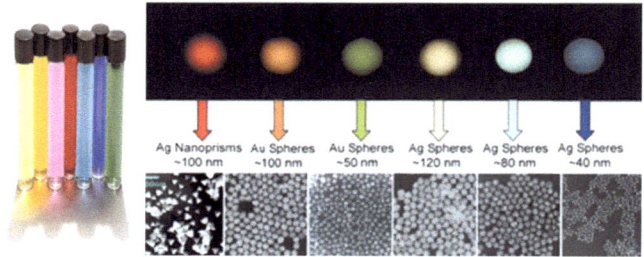

Figure 1.3: Colors of the nanoparticles with respect to their shapes and size
(Source: http://nanocomposix.com/pages/color-engineering)

In recent years, due to the unique electronic, mechanical, optical, magnetic and chemical properties of nanoparticles which are totally different from bulk of material (Wang, *et al.* 2005), nanoparticles have been matter of focused of researchers (Guzman, *et al.* 2009).These distinct and exceptional properties could be recognized to their small size and large specific surface area. For these reasons metallic nanoparticles are used in many applications in different fields of science (Guzman, *et al.* 2009).

1.4   SILVER NANOPARTICLES (AgNPs)

Silver is a naturally occurring precious metal, found as a mineral ore. It has been positioned as the 47th element in the periodic table, with atomic weight of 107.8 and having two natural isotopes 106.90 Ag and 108.90 Ag with abundance of 52 and 48% respectively (Zhou, Wang, 2012). Silver nanoparticles (AgNPs) (See figure 1.4) are of great interest due to the unique properties, (e.g. size, shape, optical, magnetic, electrical and chemical properties) that can be integrated into antimicrobial applications (Panigrahi, 2013; Szczepanowicz, *et al.* 2010; Iravani, *et al.* 2014; Landage, *et al.* 2014). The application of AgNPs as an antimicrobial agent is comparatively new (Ahmad, *et al.* 2011).

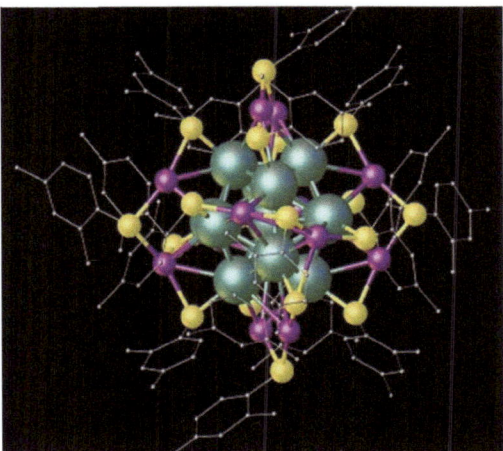

Figure 1.4: Nanoparticle structure containing 25 silver atoms (purple and green)
(Source: https://discovery.kaust.edu.sa/en/article/160/silver-nanoparticles-gold-luster)

Due to their high reactivity (because of large surface to volume ratio), AgNPs have a vital role in bacterial growth inhibition, in aqueous or solid media (Guzman, *et al.* 2009) thus are used in the medical field for antimicrobial applications such as burn treatment, prevention of bacteria colonization on catheters, eliminate microorganisms on textile fabrics (Becheri, *et al.* 2008), and as disinfectants in water treatment (Guzman, *et al.* 2009; Maiti, *et al.* 2014). The antimicrobial activity of colloid AgNPs are subjective to the dimensions of the particles *i.e.* smaller the particle size, the greater would be the antimicrobial effects (Guzman, *et al.* 2009).

Silver is commonly used in the nitrate form to persuade antimicrobial effects, but because of the use of silver nanoparticles, the availability of the surface area become increase for the microbe to be exposed to. There are many physical, biological and chemical procedures for the synthesis of silver nanoparticles (AgNPs) (Aashritha, 2013; Szczepanowicz, *et al.* 2010; Iravani, *et al.* 2014)

1.5   METHODS OF NANOPARTICLE SYNTHESIS

Various techniques for nanoparticles (nanomaterials) preparation are summarized below:

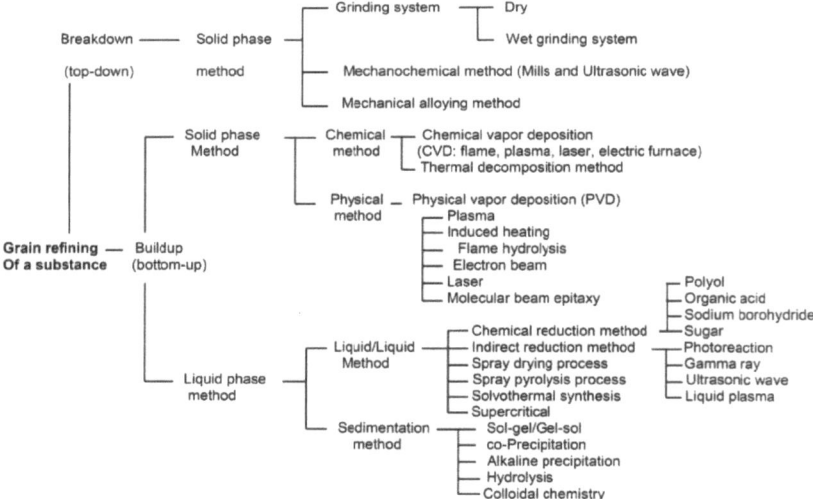

Figure 1.5:Typical synthetic methods for nanoparticles for the top-down and bottom-up approaches (Horikoshi, *et al.*2013)

From ancient time, two approaches are well known in the preparation of ultrafine particles. The first is the breakdown (top-down) method in which an external force is applied to solid that breaks it up into smaller particles. The second is the build-up (bottom-up) method that produces nanoparticles starting from atoms of gas or liquids based on atomic transformations or molecular condensations (Horikoshi, Serpone, 2013).The nanoparticles production through physical and chemical processes can be possibleby both the so called 'top-down' and 'bottom-up' procedures(see figure 1.5) (Pacioni, *et al.* 2015; Landage, *et al.* 2014). There are so many techniques for the synthesis of AgNPs, *i.e.* chemical and photo-reduction in reverse micelles, spark discharge decomposition in organic solvents and cryochemical synthesis etc. Mostly the physical and chemical procedures which are used for the production of silver nanoparticles are extremely expensive and they use toxic and hazardous chemicals which may pose potential biological and environmental risks (Aashritha, 2013; Iravani, *et al.* 2014).

Thus at the present time to synthesis the nanoparticles, there is also an increasing need to develop eco-friendly procedures, which do not use toxic chemicals in the synthesis and are economically and environmentally viable. The growing essentials to develop economically viable and environmentally friendly technologies for material synthesis led to explore the green synthesis methods include mixed valence polyoxometalates, tollens, polysaccharides, irradiation and biological methods which are beneficial over conventional methods use chemical agents involved in environmental toxicity (Iravani, *et al.* 2014).

1.5.1 CHEMICAL REDUCTION METHOD

Chemical reduction is the most commonly applied procedure for the preparation of silver nanoparticles (AgNPs) as stable, colloidal dispersion in organic solvents or water (Aashritha, 2013; Szczepanowicz, *et al.* 2010). Reductants which are frequently used involve borohydride, citrate, ascorbate and elemental hydrogen. The silver ions (Ag^+) reduction in aqueous solution usually yields colloidal silver with particle size of several nanometers in diameter.

A variety of chemical reduction procedures have been applied to synthesize stable and several shapes of silver nanoparticles (AgNPs) in water by the use of different reducing agents' i.e. ascorbic acid, dry methane, sodium borohydride ($NaBH_4$), (Figure 1.6) hydrazine and dimethyl formamide, sodium citrate, (Aashritha, 2013; Landage, *et al.* 2014), ascorbate, elemental hydrogen, polyol process, tollens reagent, N, N-dimethyl formamide (DMF), and poly (ethylene glycol)-block co-polymers are used for reduction of silver ions (Ag^+) in aqueous or non-aqueous solutions.

These reducing agents reduce Ag^+ and lead to the formation of metallic silver (Ag^0), which is followed by agglomeration into oligomeric clusters. These clusters eventually lead to the formation of metallic colloidal silver particles (Iravani, *et al.* 2014). . The reducing agent vary in the choice of, the relative concentrations and quantities of reagents, duration of reaction, temperature, along with the diameters of the nanoparticles produced. Approximately in all of them the colloidal silver products are found in turbid and greenish-yellow or brown color (D. Solomon, *et al*. 2007).

1.5.1.1 REDUCTION BY CITRATE ION

In 1951, Turkevich described the synthesis of gold nanoparticles in aqueous solution at boiling temperature by the use of sodium citrate to reduce $AuCl_4$. Since then, this methodology, known as the Turkevich's method, has been extended to other metals, like in the case of silveras shown in figure 1.6. From the pioneering studies, it is now proven that citrate worked both to reduce the metal cation and stabilize the subsequent nanoparticles. Also, it was believed that this reactant played a role on controlling the growth of the particles. Citrate controls the size and shape of AgNPs and it is found that by means of boiling method at different concentrations, AgNPs with plasmon maximum absorbance at 420 nm were synthesized (Pacioni, *et al*. 2015).

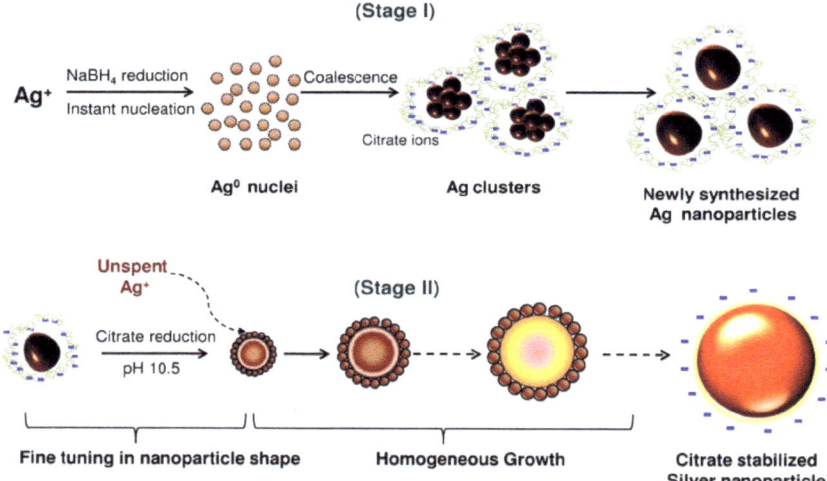

Figure 1.6: Synthesis of silver nanoparticles by chemical reduction (Agnihotri, *et al*. 2014)

1.5.2 BIO-BASED METHODS

Mostly the physical and chemical procedures, which are used for the production of silver nanoparticles, are extremely expensive and they use toxic and hazardous chemicals which may pose potential biological and environmental risks (Aashritha, 2013; Iravani, *et al*. 2014).Criteria of solvent medium and eco-friendly nontoxic reducing and stabilizing agent selection are the most essential concerns, must be considered in green synthesis of nanoparticles (NPs) (Iravani, *et al*. 2014).

Thus to synthesis the nanoparticles, the economically and environmentally viable way is inevitable. To achieve this, researchers are using biosynthesis of nanoparticles using organisms' ranges from simple prokaryotic bacterial cells to eukaryotic fungi and plants (Iravani, *et al*. 2014; Vahabi, *et al*. 2011). Some examples of nanoparticle production include using bacteria for gold, silver, cadmium, zinc, magnetite, and iron NPs; yeasts for silver, lead and cadmium NPs; fungi for gold, silver and cadmium NPs; algae for silver and gold NPs; plants for silver, gold, palladium, zinc oxide, platinum, and magnetite NPs (Iravani, *et al*. 2014), and recently the use of fungus, is a simple and viable alternative to most of complex physical and chemical synthesis methods of preparing nanoparticles (Vahabi, *et al*. 2011). The green approach of nanoparticle synthesis through biological entities has been achieving great attention over several other physicochemical methods (Devi, *et al*. 2013).

It is identified that biological entities like microorganisms and other living cells are the best machines that have operating parts at the Nano-scale level and perform numerous functions ranging from energy generation to extraction of targeted materials. The use of such microorganisms like bacteria, herbal extract, yeast and fungi in the synthesis of nanoparticles is relatively modern activity (Vahabi, *et al*. 2011).

Certain yeast, bacteria and now fungi play a vital role in remediation of toxic metals through metal ions reduction (Hussain, *et al*. 2011). For example, environmentally-friendly microorganisms could reduce the toxicity in metallic nanoparticles production process by metal ion reduction or by formation of insoluble complexes with metal ions in the form of colloidal particles. Some examples of the utilization of biological entities in the nanoparticle synthesis of different chemical compositions include the following:

i. Use of ribosomes for biosynthesis of gold nanoparticles.

ii. Use of bacteria for synthesis of cadmium sulfide, zinc sulfide, magnetite, iron sulfide and silver nanoparticles.

iii. Using the leaves, sprouts and roots of different plants for the production of nanoparticles of variable forms and morphologies.

iv. Use of yeast for the production of lead sulfide and cadmium sulfide nanoparticles.

v. Application of fungi (*i.e. Trichoderma pseudokoningii*, figure 1.7) for the synthesis of AgNPs (Vahabi, *et al*. 2011), which is also the focus of present research.

Synthesis of AgNPs through fungi has so many benefits over the above mentioned approaches. They have tolerance to high metal nanoparticles concentration in the medium, easy management of nanoparticles and good dispersion of nanomaterials. As compared to bacterial broth, fungal broth can easily be filtered by filter paper or other similar used equipment.

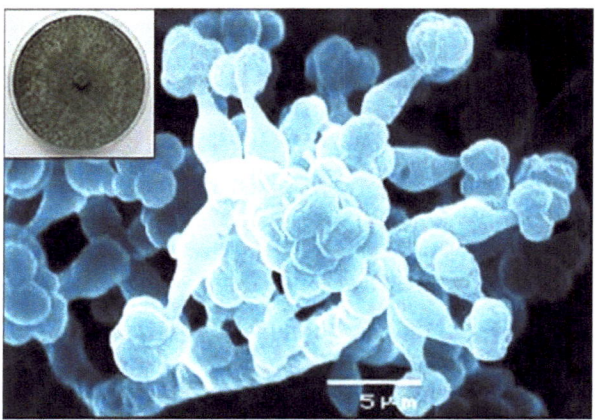

Figure 1.7: Trichoderma pseudokoningii
(Source: http://www.themushroompeople.com)

1.6 DETECTION OF SILVER NANOPARTICLES

To determine whether silver nanoparticles are actually synthesized or not, characterization of the nanoparticles is done by examining their size, shape, and quantity etc. A number of different measurement procedures and techniques can be used for this purpose, i.e. Atomic Force Microscopy (AFM), Scanning Electron Microscopy (SEM), Absorbance Spectroscopy and Dynamic Light Scattering (DLS), UV-Visible spectrophotometry, X-ray diffraction (Prema, 2011), TEM micrographs etc. (Rodríguez-León, *et al.* 2013).

1.6.1 UV-VISIBLE SPECTROPHOTOMETRY

UV-visible spectroscopy is one of the most popular characterization techniques to determine nanoparticle formation and its properties. The distinctive colors of colloidal gold and silver are because of a phenomenon known as plasmon absorbance. Incident light creates oscillations in conduction electrons on the surface of the nanoparticles and electromagnetic radiation is absorbed (D. Solomon, *et al.* 2007). Additionally, it is known that the spectrum surface plasmon resonance of nanoparticles is subjected to the size, shape, inter-particle interactions, free electron density and surrounding medium, which shows that it is an effective tool for monitoring the electron injection and aggregation of NPs. The plasmon resonance band broadens with the decrease in particle size in accordance with the quantum size theories. Likewise, the increasing integrated peak area of the band indicates decrease in inter-particle spacing, which shows aggregation (Desai, *et al.* 2012).

1.7 SILVER NANOPARTICLES AND ANTIMICROBIAL ASSAY

For a long time till present, the growth of unwanted microorganisms is a problem for the food industry and in the medical field etc. So, there is a need for methods to kill or slow down the growth of unwanted microorganisms. An interesting alternative method is the use of silver nanoparticles (Prema, 2011). The inhibitory action of silver ions and silver compounds had been historically valued and applied as a beneficial therapeutic agent for the prevention of wound infections (Singh, *etal.*2014).

Nanoparticles are considered as a feasible alternative to antibiotics and silver has always been used against various diseases; in the past it was used as an antiseptic and antimicrobial due to its low cytotoxicity. AgNPs are able to physically interact with the cell surface of various microorganisms (Franci, *et al.* 2015). Information about the structure of microbe is required, to attain understanding of this effect. Especially, the microbial membrane and the contained proteins are area of interest, because the silver has to react with it in order to penetrate the microbes. Silver is renowned for a long time because of its toxic effects against a wide range of bacteria, yeast and fungi etc. The bactericidal or fungicidal effect of silver can be classified into two groups; the reactive component may be either silver ions or silver nanoparticles. A clear distinction between ions and particles is; silver ions are charged atoms (Ag^+) while silver nanoparticles (Ag^0) are single crystals of nano size dimensions. Regardless of the fact that the bactericidal effect of silver ions is renowned and useful but it is not fully understood still. Experiments have proved that silver ions are capable to cause structural changes in the cell membrane.

The membrane of bacteria consists of lots of sulfate-containing enzymes. This inactivation makes the membrane vulnerable and easier to penetrate for silver ions. Silver ions continue damage different parts and functions of the cell by interacting with sulfate-groups, inside the cell, which are often located in the active site of enzymes. This interaction of silver ions with the active site produces an inactivation of the enzymes. Silver ions are also capable to interact with phosphorus-groups of molecules and certain experiments proved that causes severe effects. The interaction between silver ions and the backbone of DNA, as a result the bacterium became unable to replicate itself or transcribes mRNA for new proteins. All these changes slow down the growth of the bacteria and finally kill it (Figure 1.8).The bactericidal mechanism of silver nanoparticles on microorganism is still almost unknown. It has been proposed that the effect that produced is relatively same of the mechanisms that caused the bactericidal effect of silver ions (Prema, 2011).

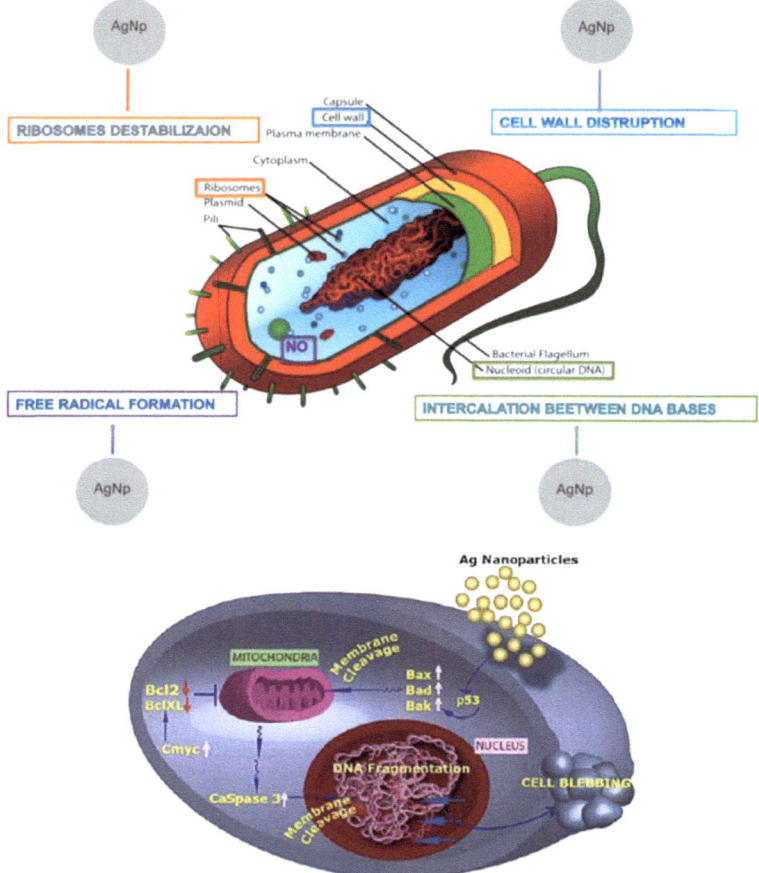

Figure 1.8: Antimicrobial mechanisms of silver nanoparticles
(Franchi, *et al.* 2015; Gopinath, *et al.* 2010)

Numerous researches shows that silver nanoparticles interact into the cell membrane of microbes interrupting their metabolisms i.e. permeability and respiration and in several research works it is realized that smaller particles have large surface area and are more effective than bulk particles against pathogenic microorganisms (Balu, *et al.* 2012).

CHAPTER 2
AIM OF THE RESEARCH

Nanotechnology has become one of the most prominent and exciting forefront areas in chemistry, physics, biology and engineering. It presents great promise for providing us in the near future with many break through that will change the direction of technological advances in wide range of applications. Nanoparticles of noble metals are of great interest todays due to their possible applications in various fields of science.

A variety of preparation techniques have been reported for the synthesis of AgNPs. Chemical reduction is the most frequently applied method for the preparation of silver nanoparticles as stable, colloidal dispersions in water or organic solvents. On the other hand, the extracts from bio-organisms also act both as reducing and capping agents in the synthesis of silver nanoparticles.

The reduction of Ag+ ions by combinations of biomolecules found in these extracts such as enzymes/proteins, amino acids, polysaccharides, and vitamins is environment friendly, yet chemically complex (Landage, *et al*. 2014). The use of silver and other metal ions for their sustained anti-fungal, anti-bacterial and anti-viral effects have been practiced for a long time. Such effects are generally referred as oligodynamic action. Silver ion has been known to be effective against a broad range of microorganisms.

This study aimed to:

a).   To synthesize silver nanoparticles (AgNPs) by the use of simple, effective environmental friendly and economically feasible, chemical reduction method and biosynthesis of nanoparticles by using specie of a fungus *Trichoderma pseudokoningii*.

b).   To evaluate the antifungal activity of synthesized AgNPs against a yeast species *Candida albicans* and antibacterial activity on two bacteria, *Escherichia coli* and *Staphylococcus aureus*.

CHAPTER 3
MATERIALS AND METHODS

In present study, two methods were applied for the synthesis of nanoparticles, chemical reduction (Turkevich method) and biosynthesis of silver nanoparticles by using fungus (*Trichoderma pseudokoningii*).

3.1 MATERIALS

3.1.1 REAGENTS

Silver nitrate ($AgNO_3$), Citrate of sodium, Glucose, Peptone, Agar, Nutrient agar and Double-distilled deionized water.

3.1.2 SOURCE OF MICRO-ORGANISMS

All microorganism cultures, *Trichoderma pseudokoningii*, *Candida albicans*, *Escherichia coli*, *Staphylococcus aureus* were obtained from the culture collections of Microbiology Department, Federal Urdu University Arts, Science and Technology, Karachi, Pakistan.

3.1.3 COMPOSITION OF SDB (Sabouraud Dextrose Broth)

Glucose	4%
Peptone	1%
Deionized water	100 mL

3.1.4 CHEMICAL COMPOSITION OF SDA (Sabouraud Dextrose Agar)

Glucose	4%
Agar	2%
Peptone	1%
Distilled water	100 mL

3.1.5 CHEMICAL COMPOSITION OF NUTRIENT AGAR

Nutrient agar	4%
Agar	1%
Distilled water	100 mL

3.2 SYNTHESIS OF SILVER NANOPARTICLES (AgNPs)

3.2.1 CHEMICAL REDUCTION METHOD (TURKEVICH METHOD)

Turkevich method was used for the synthesis of AgNPs, through the reduction of silver nitrate ($AgNO_3$) with Citrate of sodium (sigma aldrich) as a reducing agent. Double-distilled deionized water was used as solvent. Aqueous solution of silver nitrate (1.0 mM and 6.0 mM) and aqueous solution of sodium citrate (1.0 mM and 2.0 mM) was mixed by 1:1 ratio and was kept at boiling temperature for 4 to 6 minutes till the solution turned amber yellow (Aashritha, 2013),as shown in figure 3.1, indicating the formation of colloidal silver nanoparticles.

Figure 3.1: Chemical Synthesis of silver nanoparticles

3.2.2 BIOLOGICAL METHOD (BY FUNGUS *TRICHODERMA PSEUDOKONINGII*)

3.2.2.1 PRODUCTION OF BIOMASS

Sabouraud Dextrose Broth (SDB) was used for the qualitative cultivation of dermatophytes, yeast, molds and aciduric microorganisms. 4 gm of glucose and 1 gm of peptone were added in 100 ml distilled water followed by stirring at a stable temperature to dissolve completely. The mixture was sterilized by autoclaving at 121°C for 20 minutes. 8 to 10 scoops of fungus (*Trichoderma pseudokoningii*) were put into SDB and the culture flasks were incubated on an orbital shaker at room temperature and agitated at 150 rpm.

This biomass was harvested after 72 hours of growth by filtering through a filter paper followed by extensive washing with sterile double-distilled deionized water to remove any medium contents from the biomass(Figure 3.2(b)).

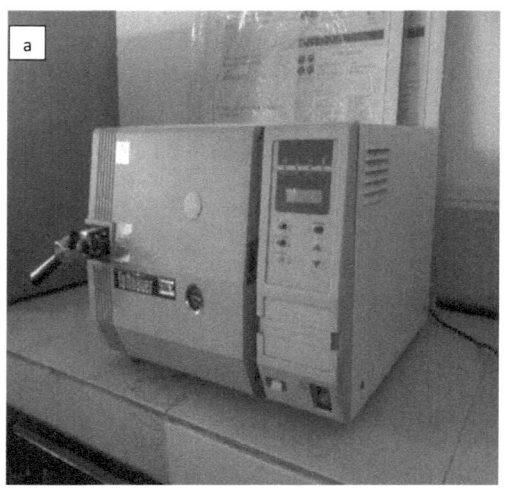

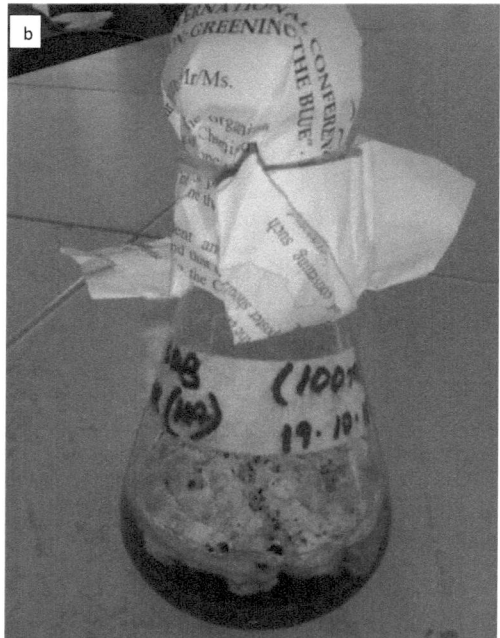

Figure 3.2: (a) Autoclave for sterilizing; (b). Biomass of fungus *Trichoderma pseudokoningii*

3.2.2.2 BIOSYNTHESIS OF SILVER NANOPARTICLES (AgNPs)

Approximately 21 gm of biomass (wet weight) was brought into contact with 100 mL sterile double-distilled deionized water for 48 hours at room temperature in a flask. After incubation, cell filtrate was filtered by filter paper and 0.017 gm of $AgNO_3$ was added into the filtrate to yield Ag+ ions, and reaction was carried out under dark environment.

After 72 hours of reaction, the solution turned yellow to brown (Figure 3.4), the change in color, indicating the formation of silver nanoparticles.

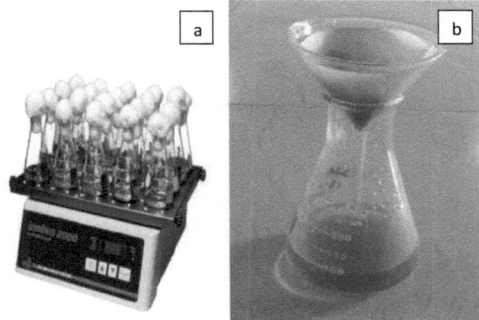

Figure 3.3: (a) shaker; (b) Filtration of fungus

Figure 3.4: Conversion of pale yellow to brown color of biosynthesized AgNPs by fungus

3.3 CHARACTERIZATION OF SILVER NANOPARTICLES

The chemical reduction and bio-reduction of Ag ions in aqueous solutions were monitored by UV-Vis spectra of the solution between 350 nm and 750 nm using spectrophotometer (Jenway 6310) as shown in Figure 3.5. The double-distilled deionized water used as a blank reference.

Figure 3.5: Spectrophotometer (Jenway 6310)

3.4 ANTIMICROBIAL ASSAYS

3.4.1 PREPARATION OF SDA PLATES FOR ANTIFUNGAL ASSAY

SDA is a selective medium primarily used for the isolation of dermatophytes, other fungi and yeasts but can also grow filamentous bacteria such as Nocardia. The acidic pH of this medium (pH about 5.0) inhibits the growth of bacteria but permits the growth of yeasts and most filamentous fungi. 4 gm of glucose 2 gm of agar and 1 gm peptone were added in 100 ml distilled water followed by stirring at a stable temperature to dissolve completely.

 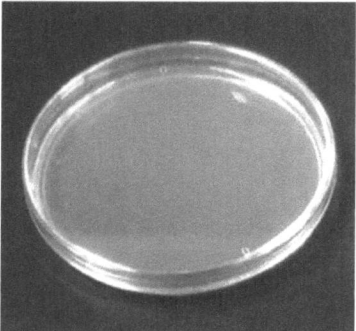

Figure 3.6: Sabouraud Dextrose Agar (SDA) Plates

The mixture was sterilized by autoclaving at 121°C for 20 minutes. Mix well and pour in to sterile plates and left few minutes to solidify, as shown in figure 3.6. Introduce *Candida albicans* in SDA plates and analyzed antifungal activity of silver nanoparticles (which are synthesized by chemical reduction method and biosynthesized by fugal method) against *Candida albicans* by bore well method. These agar plates were incubated for 48 hours at 37°C ± 2°C, then a significant range of zones of inhibition of fungus were visualized, as shown in Figure 4.4 (Chapter 4).

3.4.2 PREPARATION OF NUTRIENT AGAR PLATES FOR ANTIBACTERIAL ASSAY

Nutrient Agar is a general purpose, nutrient medium used for the cultivation of microbes supporting growth of a wide range of non-fastidious organisms. Nutrient agar is popular because it can grow a variety of types of bacteria and fungi, and contains many nutrients needed for the bacterial growth. 4 gm of nutrient agar and 1 gm agar were added in 100 ml distilled water followed by stirring at a stable temperature to dissolve completely.

The mixture was sterilized by autoclaving at 121°C for 20 minutes. Mix well and pour in to sterile plates and left few minutes to solidify (Figure 3.7). In nutrient agar plates *Escherichia coli* and *Staphylococcus aureus* were introduced for antibacterial activity. Analysis of silver nanoparticles' antibacterial activity was performed against these bacteria by bore well method. A significant range of zones of bacterial growth inhibition were visualized on agar plates after the incubation of 48 hours at 37°C ± 2°C. These results indicated that the synthesized silver nanoparticles (AgNPs) have stronger antimicrobial assay as shown in figure 4.5 (Chapter 4).

Figure 3.7: Nutrient Agar Plates

CHAPTER 4
RESULTS AND DISCUSSION

AgNPs were synthesized by using chemical reduction method and by the use of fungus, *Trichoderma pseudokoningii* as described in the previous chapter.

4.1 SYNTHESIS OF AgNPs BY CHEMICAL REDUCTION METHOD

Turkevich method was used for the synthesis of $AgNO_3$. The colloidal solution turned pale yellow at a uniform temperature shown that the silver nanoparticles were prepared by the chemical reduction method as shown in figure 3.1 Chapter 3. After the reduction of aqueous silver salts with sodium citrate within duration of 5-8 minutes the silver nanoparticles are synthesized, after the completion of reaction color change appear, it is renowned that, the silver nanoparticles indicates pale yellow color (Balu, *et al.* 2012). Silver nanoparticles are well-known to exhibit a characteristic surface plasmon resonance band that can be measured by UV-Vis spectroscopy.

4.1.1 OPTICAL SPECTROSCOPY MEASUREMENTS

Optical spectroscopy is extensively practiced for the characterization of nanomaterials. To characterize the silver nanoparticles we have produced, UV-Visible spectroscopy technique is used. UV-Vis spectroscopy, the one of the most extensively used technique for structural characterization of nanoparticles. It is usually accepted that UV-Vis spectroscopy could be used to study size and shape-controlled nanoparticles in aqueous solution (Hussain, *et al.* 2011).

UV-Vis spectroscopy is used to examine the absorption spectra (Appendix 1 and 2) of both samples of silver nanoparticles prepared by Turkevich method, silver nanoparticles presented surface plasmon resonance in the range 450 nm and peak boarding shows that the particles are poly dispersed sharp peak defines the uniformity of particles size. The shape and position of the plasmon absorption depends on the particles size, shape and the dielectric constant of the surrounding medium (Guzman, *et al.* 2009).

In this study the size of metal nanoparticles can be predicated based on its color observation. UV-Vis spectroscopy indicates the plasmon band of the silver nanoparticle suspensions, indicating a distinctive absorbance peak for nanoparticles centered at 450 nm as shown in figure 4.1 and 4.2. The symmetrical shape of the plasmon band can show a relative sharp particle size distribution (Aashritha, 2013).

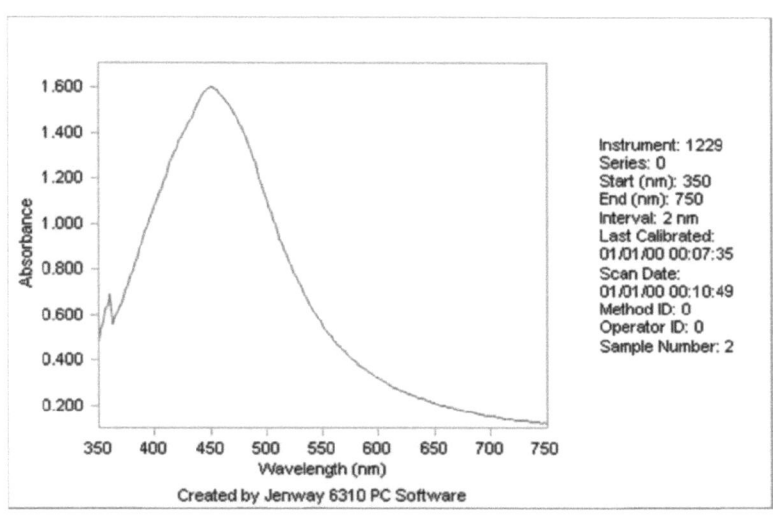

Figure 4.1: UV-Vis absorption spectrum of silver nanoparticles prepared using chemical synthesis (1.0 mM aqueous solution of AgNO3 and 1.0 mM aqueous solution of sodium citrate)

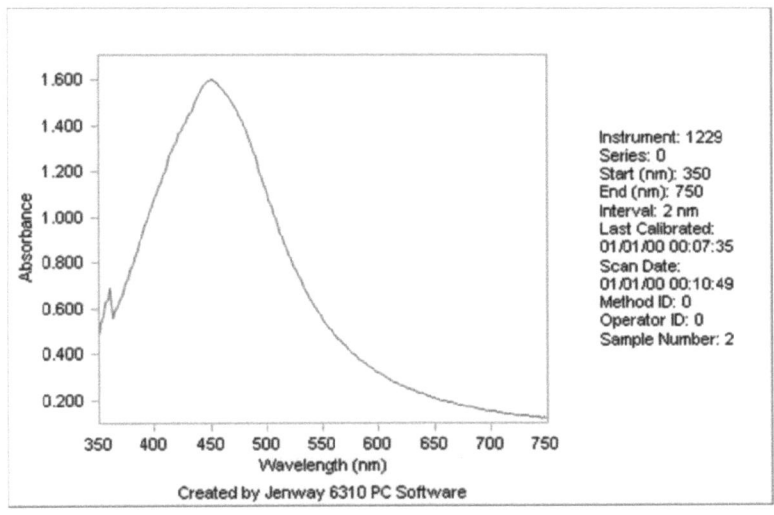

Figure 4.2: UV-Vis absorption spectrum of silver nanoparticles prepared using chemical synthesis (6.0 mM sol. of AgNO3 and 2.0 mM sol. of sodium citrate)

4.2 BIOSYNTHESIS OF SILVER NANOPARTICLES BY FUNGUS (*TRICHODERMA PSEUDOKONINGII*)

Nowadays, the green approach for the synthesis of silver nanoparticles by the use of biological entities has been achieving a profound interest over many other physicochemical procedures which have many disadvantages. Certain microorganisms play a vital role in remediation of toxic metals through reduction of the metal ions, in other ways they are not toxic. Because of the wide variety of the advantages over yeast, bacteria, actinomycetes, plants and certain other physical and chemical processes, fungi become the best choice for nanotechnologists (Devi, *et al.* 2013).

In this work, silver nanoparticles (AgNPs) are also biosynthesized with the use of fungus *Trichoderma pseudokoningii*. The addition of fungal extract to silver nitrate ($AgNO_3$) solution results in the color change of solution. It can be observed that the prior pale yellow color of the reaction solution is converted into the brownish color after 72 hours of reaction. The appearance of a yellowish-brown color in mixture containing the biomass extract is a clear sign of the formation of silver nanoparticles in the reaction mixture (Figure 3.4). The color of the reaction solution is because of the excitation of surface plasmon vibrations (essentially the vibration of the group conduction electrons) in the silver nanoparticles (Vahabi, *et al.* 2011).

4.2.1 OPTICAL SPECTROSCOPY MEASUREMENTS

Optical spectroscopy is extensively practiced for the characterization of nanomaterials. To characterize the silver nanoparticles we have produced biologically, UV-Visible spectroscopy technique is used. We use UV-Vis spectroscopy to follow up with the reaction process. UV-Visible spectra recorded after 72 hours of the reaction. UV-visible absorption spectra (Appendix 3) for AgNPs prepared from *Trichoderma pseudokoningii* using silver nitrate ($AgNO_3$).

The spectra recorded from the AgNPs solution, showed an absorbance peak at 420 nm, as shown in Figure 4.3 which was specific for the silver nanoparticles. The result indicated that the high Plasmon band was observed at 420 nm after 72 h.

It was observed that the absorbance peak was centered near 420 nm, indicating the conversion of silver nitrate into silver nanoparticles.

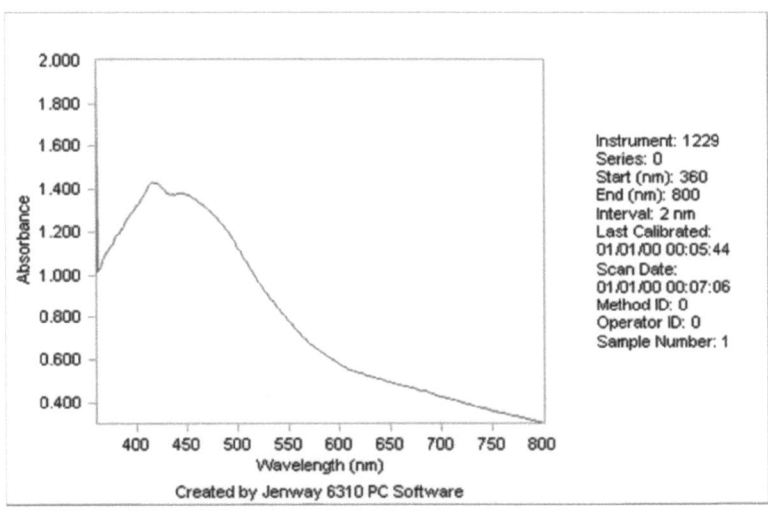

Figure 4.3: UV-Vis absorption spectrum of silver nanoparticles prepared using biological synthesis

4.3 ANTIMICROBIAL ASSAY

Nanoparticle technology is being integrated into many areas of molecular science and biomedicine. Because nanoparticles are small enough to enter nearly all parts of the body, including the circulatory system and cells in all living creatures, they have been and continue to be exploited for basic biomedical research along with clinical diagnostic and therapeutic applications. To observe the coating mechanism, it is very supportive to work with particles which are uniform in size and shape (Kekutia, *et al*. 2014).

Formerly studies reported that the antimicrobial activities of silver nanoparticles (AgNPs) are size-dependent, the smallest sized indicated the strongest effect. This research work aims leading to cheap, effective, and safe methods of nanoparticle synthesis necessary for antimicrobial activity.

Silver nanoparticles synthesized by chemical method and by fungal extract have been found highly toxic against pathogenic microorganisms. Herein, the chemically and biologically synthesized silver nanoparticles displayed antimicrobial activity against *Candida albicans*, *Escherichia coli* and *Staphylococcus aureus* as it shown clear inhibition zone. Numerous studies reveals that silver nanoparticles interact into the microorganisms disturbing their metabolic activities i.e. permeability and respiration functions and the several studies define that smaller particles have large surface area, are more effective then bulk particles towards microorganisms (Balu, *et al.* 2012). Analysis and results of the study proves that, silver nanoparticles having antimicrobial activities towards various microorganisms.

4.3.1 ANTIFUNGAL ASSAY OF SILVER NANOPARTICLES

Silver nanoparticles were analyzed for their antifungal activity against *Candida albicans*, by bore well method. *Candida albicans*, indicates the presence of certain level inhibited growth at significant measurement by the zone of inhibition (Figure 4.4).

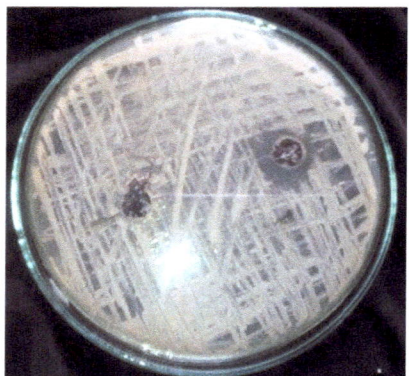

Figure 4.4: Zone of inhibition against *Candida albicans*

4.3.2 ANTIBBACTERIAL ASSAYS OF SILVER NANOPARTICLES

It was also observed that bacterial growth of *Escherichia coli* and *Staphylococcus aureus* were also independent on silver nanoparticles (AgNPs) concentration. Silver nanoparticles also inhibited the visible growth of *E. coli*, indicated by the clearly visible zone of inhibition. Likewise, *S. aureus* also showed a meaningful range of zone of inhibition (Figure 4.5), expressing the anti-microbial assay of chemically and biologically synthesized silver nanoparticles (AgNPs).

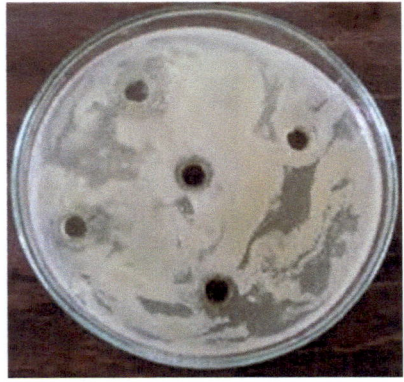 

Figure 4.5: Zone of inhibition against *Staphylococcus aureus* and *Escherichia coli*

Because of the nano size of the metallic particles it is ensures that a considerably large surface area of the particles is in contact with the microbial cells. Such a large contact surface is supposed to enhance the extant of microbial elimination. The synthesis and characterization of nano-scaled materials in term of novel physicochemical properties is of great importance in the formulation of fungicidal or bactericidal materials.

The degree of inhibition depends on the concentration of the silver nanoparticles as well as on the initial microbial population. The inhibitory action mechanism of silver ions on microorganisms' shows that upon Ag^+ treatment, DNA loses its replication capability and expression of ribosomal subunit proteins, along with other cellular proteins and enzymes necessary to ATP production, becomes inactivated.

It has also been hypothesized that due to Ag^+, the function of membrane bound enzymes, in the respiratory chain are primarily affected (Aashritha, 2013). Finally, the antimicrobial susceptibility of silver nanoparticles synthesized was investigated. Antimicrobial activities of the silver colloidal sols were assessed using the bore well method, leading to inhibition of microbial growth.

CHAPTER 5
CONCLUSION

AgNPs have gained considerable interest because of their unique properties, and proven applicability in diverse areas such as medicine, catalysis, textile engineering, biotechnology, Nano biotechnology, bioengineering sciences, electronics, optics, and water treatment. These NPs have significant inhibitory effects against microbial pathogens, and are widely used as antimicrobial agents in a diverse range of products.

In this study, simple, fast and eco-friendly methods for synthesis of AgNPs with antimicrobial activity were demonstrated. Chemically, AgNPs were synthesized successfully by chemical reduction of $AgNO_3$. Nanoparticles reduced with sodium citrate while biologically; synthesis was done by the use of fungus *Trichoderma pseudokoningii*, showed high activity against certain microorganisms.

Color changes occur due to surface plasmon resonance during the reaction with the ingredients present in the fungal extract resulting in the formation of silver nanoparticles biologically and chemically due to the reducing agent, which is confirmed by UV–Vis spectroscopy. The nanoparticles were characterized by UV-Vis spectra show the characteristic plasmon absorption peak for the silver nanoparticles ranging from 420 to 450 nm. The UV-Vis absorption spectra of silver particles confirmed the Nano crystalline character of the particles synthesized using various methods. The silver nanoparticles synthesized by these methods are strong candidates for its use in biological systems.

These chemically and biologically synthesized AgNPs were then used to evaluate antimicrobial activity against *Candida albicans*, *Staphylococcus aureus* and *Escherichia coli*. The antibacterial activity of the nanoparticles dispersion was measured by bore well method. The antibacterial activity of AgNPs was apparent from the zone of inhibition. The antimicrobial assay showed that silver nanoparticles presented good antibacterial performance against bacteria and fungus.

The synthesis of nanoparticles using these techniques is simple and low-cost approaches. The processes are useful in preparation of antimicrobial agents and could be exploited for further biomedical application. The results of this study clearly demonstrated that the colloidal silver nanoparticles inhibited the growth and multiplication of the tested bacteria and fungi, including *S. aureus*, *E. coli* and *C. albicans*. This research work leads to cheap, effective and safe methods of nanoparticle synthesis, necessary for antimicrobial activities.

This work also integrates nanotechnology and microbiology, leading to possible advances in the formulation of new types of microbicides. However, future studies on the biocidal influence of this nanomaterial on these microorganisms are necessary in order to fully evaluate its possible use as a new fungicidal or bactericidal material.

BIBLIOGRAPHY

Aashritha, S. 2013. Synthesis of silver nanoparticles by chemical reduction method and their antifungal activity. Int. Res. J. Pharm. 4(10)

Agnihotri, S., Mukherji, S., &Mukherji, S. (2014). Size-controlled silver nanoparticles synthesized over the range 5–100 nm using the same protocol and their antibacterial efficacy. RSC Advances, 4(8), 3974-3983.

Ahmad, M.B., Lim, J.J., et al 2011. Synthesis of Silver Nanoparticles in Chitosan, Gelatin and Chitosan/Gelatin Bionanocomposites by a Chemical Reducing Agent and Their Characterization. Molecules. 16, 7237-7248

Awwad, A.M., Salem, N.M., Abdeen, A.O. 2013. Green synthesis of silver nanoparticles using carob leaf extract and its antibacterial activity. Int. J. Ind. Chem. (IJIC). 4:29

Balu, S.S., Bhakat, C., Harke, S. 2012. Synthesis of silver nanoparticles by chemical reduction and their antimicrobial activity. Int. J. Eng. Res. Tech. (IJERT). ISSN: 2278-0181.

Becheri, A., Durr, M., Lo Nostro, P., Baglioni, P. 2008. Synthesis and characterization of zinc oxide nanoparticles: application to textiles as UV-absorbers. J. Nanopart. Res. 10: 679-689

Devi, T.P., Kulanthaivel, S., et al 2013. Biosynthesis of silver nanoparticles from Trichoderma species. Ind. J. Exp. Bio. 51: 543-547

Fortina, P., Kricka, L. J., Surrey, S., &Grodzinski, P. (2005). Nanobiotechnology: the promise and reality of new approaches to molecular recognition. TRENDS in Biotechnology, 23(4), 168-173.

Franci, G., Falanga, A., Galdiero, S., Palomba, L., Rai, M., Morelli, G., Galdiero, M. 2015. Silver Nanoparticles as Potential Antibacterial Agents. Molecules. 20: 8856-8874

Gherbawy, Y.A., Shalaby, I.M., El-sadek, M.S.A., Elhariry, H.M., Banaja, A.A. 2013. The Anti-Fasciolasis properties of silver nanoparticles produced by Trichodermaharzianumand their improvement of the Anti-Fasciolasis drug Triclabendazole. Int. J. Mol. Sci. 14, 21887-21898; doi:10.3390/ijms141121887

Gopinath, P., Gogoi, S. K., Sanpui, P., Paul, A., Chattopadhyay, A., &Ghosh, S. S. (2010). Signaling gene cascade in silver nanoparticle induced apoptosis. Colloids and Surfaces B: Biointerfaces, 77(2), 240-245.

Guzman, M.G., Dille, J., Godet, S. 2009. Synthesis of silver nanoparticles by chemical reduction method and their antibacterial activity. Int. J. Chem. Biomol. Eng. 2:3

Horikoshi, S., Serpone, N. 2013. Microwaves in Nanoparticles Synthesis. Wiley-VCH Verlag GmbH & Co. KGaA.

Hussain, J.I., Kumar, S., Hashmi, A.A., Khan, Z. 2011. Silver nanoparticles: preparation, characterization, and kinetics. Adv. Mat. Lett. 2(3): 188-194

Iravani, S., Korbekandi, H., Mirmohammadi, S.V., Zolfaghari, B. 2014. Synthesis of silver nanoparticles: chemical, physical and biological methods. Res. Phar. Sc. 9(6): 385-406

Kamikawa, Y., Hirabayashi, D., Nagayama, T., Fujisaki, J., Hamada, T., Sakamoto, R., Kamikawa, Y., Sugihara, K. 2014. In Vitro Antifungal Activity against Oral Candida Species Using a Denture Base Coated with Silver Nanoparticles. J. Nanomat.

Kekutia, S., Saneblidze, L., Mikelashvili, V., Markhulia, J., Tatarashvili, R., Daraselia, D., Japaridze, D. 2014. A new method for the synthesis of nanoparticles for biomedical applications. Eur. Chem. Bull. 4(1): 33-36

Kheybari, S., Samadi, N., Hosseini, S.V., Fazeli, A., Fazeli, M.R. 2010. Synthesis and antimicrobial effects of silver nanoparticles produced by chemical reduction method. DARU. J. Pharm. Sc. 18(3): 168–172

Kim.,Soo-Hwan., Lee, H.S., Ryu, D.S., Choi, S.J., Lee, D.S. 2011. Antibacterial Activity of Silver-nanoparticles Against *Staphylococcus aureus* and *Escherichia coli*. 39:77-85

Landage, S.M., Wasif, A. I., Dhuppe P. 2014. Synthesis of nanosilver using chemical reduction methods. Int. J. Adv. Res. Eng. App. Sc. ISSN: 2278-6252

Maiti, S., Krishnan, D., Barman, G., Ghosh, S.K., Laha, J.K. 2014. Antimicrobial activities of silver nanoparticles synthesized from Lycopersiconesculentum extract. J. Anal. Sci. & Tech. 5:40

Mani, U., Dhanasingh, S., Arunachalam, R., Paul, E., Shanmugam, P., Rose, C., Mandal, A.B. A simple and green method for the synthesis of silver nanoparticles using RicinusCommunis Leaf extract. Prog. Nanotech. Nanomat. 2

Mallmann, E.J., Cunha, F.A., N.M.F. Castro, B., Maciel, A.M., Menezes, E.A., Fechine, P.B.A. 2015. Antifungal activity of silver nanoparticles obtained by green synthesis. Rev. Inst. Med. Trop. Sao. Paulo. 57(2): 165–167

Morais, M. G. D., Martins, V. G., Steffens, D., Pranke, P., & da Costa, J. A. V. (2014). Biological applications of nanobiotechnology. Journal of nanoscience and nanotechnology, 14(1), 1007-1017.

Nagarajan, R., & Hatton, T. A. (2008). Nanoparticles: synthesis, stabilization, passivation, and functionalization (Vol. 996, pp. 2-14). Washington DC:: American Chemical Society.

Pacioni, N., D. Borsarelli, C., Rey, V., V. Veglia, A. Spr. Int. Pub. Swit.

Panigrahi, T. 2013. Synthesis and characterization of silver nanoparticles using leaf extract of *Azadirachtaindica*. M.Sc. Thesis. Life Sc. Dept. Nat. Ins. Tech. India.

Prema, P. 2011. Chemical Mediated Synthesis of Silver Nanoparticles and its Potential Antibacterial Application. Progress in Mole. Env. Bioeng. From Ana. Mod. To Tech. App.

Rajeshkumar, S., Malarkodi, C. 2014. In Vitro Antibacterial Activity and Mechanism of Silver Nanoparticles against Foodborne Pathogens. Bioinorg. Chem. App.

Rodríguez-León, E., Iñiguez, R., Navarro, R.E., Herrera-Urbina, R., Tánori, J., Iñiguez-Palomares, C., Maldonado, A. 2013. Synthesis of silver nanoparticles using reducing agents obtained from natural sources (Rumexhymenosepalus extracts). Nanosc. Res. Lett. 8:318

Singh, K., Panghal, M., Kadyan, S., Chaudhary, U., Yadav, J.P. 2014. Antibacterial Activity of Synthesized Silver Nanoparticles from Tinosporacordifolia against Multi Drug Resistant Strains of Pseudomonas aeruginosa Isolated from Burn Patients. J. Nanomed. Nanotechnol. 5:2

Solomon, S., Bahadory, M., et al. 2007. Synthesis and Study of Silver Nanoparticles. J. Chem. Edu. 84

Szczepanowicz, K.P., Stefanska, J., Socha, R.P., Warszynski, P. 2010. Preparation of silver nanoparticles via reduction and antimicrobial activity. Physicochem. Probl. Miner. Process. 45: 85-98

Taylor, R.A., Coulombe, S., Otanicar, T., Phelan, P.E., Gunawan, A., Lv, W., Gary, R., Ravi, P., Tyagi, H. 2013. Small particles, big impacts: A review of diverse applications of nanofluids. J. App. Phys. 113: 011301.

Tsuzuki, T. (2009). Commercial scale production of inorganic nanoparticles.International journal of nanotechnology, 6(5-6), 567-578.

Umer, A., Naveed, S., Ramzan, N., Rafique, M.S., Imran, M. 2014. A green method for the synthesis of Copper nanoparticles using L-ascorbic acid. Materia (Rio. J.). ISSN 1517-7076

Vahabi, K., Mansoori, G.A., Karimi, S. 2011. Biosynthesis of silver nanoparticles by fungus Trichodermareesei. Insci. J. 1(1): 65-79; doi: 10.5640/insc.010165

Wang, H., Qiao, X., Chen, J., Ding, S. 2005. Preparation of silver nanoparticles by chemical reduction method. Coll. Sur. A: Phy. Chem. Eng. Asp. 256: 111-115

Zhou, G., Wang, W. 2012. Synthesis of Silver Nanoparticles and their Antiproliferationagainst Human Lung Cancer Cells In vitro. Ori. J. Chem. Vol. 28(2): 651-655

APPENDIX 1

UV-VISIBLE ABSORPTION SPECTRUM OF SILVER NANOPARTICLES PREPARED USING CHEMICAL SYNTHESIS

(1.0 mM AQUEOUS SOLUTION OF $AgNO_3$ AND 1.0 mM AQUEOUS SOLUTION OF SODIUM CITRATE)

wavelength	Absorbance
350	0.482
352	0.535
354	0.570
356	0.622
358	0.647
360	0.692
362	0.557
364	0.586
366	0.609
368	0.632
370	0.656
372	0.686
374	0.711
376	0.739
378	0.770
380	0.797
382	0.822
384	0.854
386	0.886
388	0.915
390	0.942
392	0.964
394	0.989
396	1.015
398	1.051
400	1.078
402	1.102
404	1.131
406	1.157
408	1.179
410	1.209
412	1.241
414	1.264
416	1.295
418	1.313
420	1.334
422	1.364
424	1.380
426	1.400
428	1.416
430	1.438
432	1.451
434	1.464
436	1.489
438	1.524
440	1.535
442	1.563
444	1.575
446	1.586
448	1.593
450	1.602
452	1.601
454	1.597
456	1.587
458	1.577
460	1.564
462	1.556
464	1.539
466	1.530
468	1.512
470	1.498
472	1.481
474	1.461
476	1.444
478	1.425
480	1.402
482	1.377
484	1.351
486	1.320
488	1.296
490	1.266
492	1.234
494	1.202
496	1.170
498	1.141
500	1.109
502	1.078
504	1.048
506	1.020
508	0.992
510	0.963
512	0.936
514	0.910
516	0.885
518	0.861
520	0.836
522	0.813
524	0.790
526	0.769
528	0.747
530	0.728
532	0.707
534	0.687
536	0.668
538	0.651
540	0.633
542	0.616
544	0.600
546	0.585
548	0.570
550	0.555
552	0.541
554	0.528
556	0.515
558	0.504
560	0.490
562	0.481
564	0.469
566	0.458
568	0.447
570	0.438
572	0.429
574	0.419
576	0.410
578	0.400
580	0.392

582	0.384	640	0.228	698	0.151
584	0.375	642	0.222	700	0.151
586	0.368	644	0.220	702	0.150
588	0.361	646	0.216	704	0.150
590	0.354	648	0.214	706	0.148
592	0.346	650	0.211	708	0.146
594	0.340	652	0.207	710	0.145
596	0.333	654	0.203	712	0.142
598	0.326	656	0.200	714	0.142
600	0.320	658	0.198	716	0.140
602	0.314	660	0.198	718	0.140
604	0.308	662	0.195	720	0.139
606	0.302	664	0.192	722	0.136
608	0.296	666	0.189	724	0.136
610	0.291	668	0.186	726	0.135
612	0.286	670	0.184	728	0.133
614	0.280	672	0.181	730	0.132
616	0.276	674	0.180	732	0.132
618	0.271	676	0.177	734	0.130
620	0.266	678	0.174	736	0.127
622	0.262	680	0.173	738	0.127
624	0.259	682	0.171	740	0.126
626	0.253	684	0.170	742	0.124
628	0.250	686	0.166	744	0.125
630	0.247	688	0.165	746	0.122
632	0.243	690	0.162	748	0.123
634	0.240	692	0.160	750	0.121
636	0.234	694	0.159		
638	0.231	696	0.155		

APPENDIX 2

UV-VIS ABSORPTION SPECTRUM OF SILVER NANOPARTICLES PREPARED USING CHEMICAL SYNTHESIS

(6.0 mM AQUEOUS SOLUTION OF $AgNO_3$ AND 2.0 Mm AQUEOUS SOLUTION OF SODIUM CITRATE)

Wavelength	Absorbance	Wavelength	Absorbance	Wavelength	Absorbance
350	0.482	426	1.400	504	1.048
352	0.535	428	1.416	506	1.020
354	0.570	430	1.438	508	0.992
356	0.622	432	1.451	510	0.963
358	0.647	434	1.464	512	0.936
360	0.692	436	1.489	514	0.910
362	0.557	438	1.524	516	0.885
364	0.586	440	1.535	518	0.861
366	0.609	442	1.563	520	0.836
368	0.632	444	1.575	522	0.813
370	0.656	446	1.586	524	0.790
372	0.686	448	1.593	526	0.769
374	0.711	450	1.602	528	0.747
376	0.739	452	1.601	530	0.728
378	0.770	454	1.597	532	0.707
380	0.797	456	1.587	534	0.687
382	0.822	458	1.577	536	0.668
384	0.854	460	1.564	538	0.651
386	0.886	462	1.556	540	0.633
388	0.915	464	1.539	542	0.616
390	0.942	466	1.530	544	0.600
392	0.964	468	1.512	546	0.585
394	0.989	470	1.498	548	0.570
396	1.015	472	1.481	550	0.555
398	1.051	474	1.461	552	0.541
400	1.078	476	1.444	554	0.528
402	1.102	478	1.425	556	0.515
404	1.131	480	1.402	558	0.504
406	1.157	482	1.377	560	0.490
408	1.179	484	1.351	562	0.481
410	1.209	486	1.320	564	0.469
412	1.241	488	1.296	566	0.458
414	1.264	490	1.266	568	0.447
416	1.295	492	1.234	570	0.438
418	1.313	494	1.202	572	0.429
420	1.334	496	1.170	574	0.419
422	1.364	498	1.141	576	0.410
424	1.380	500	1.109	578	0.400
		502	1.078	580	0.392

582	0.384	640	0.228	698	0.151
584	0.375	642	0.222	700	0.151
586	0.368	644	0.220	702	0.150
588	0.361	646	0.216	704	0.150
590	0.354	648	0.214	706	0.148
592	0.346	650	0.211	708	0.146
594	0.340	652	0.207	710	0.145
596	0.333	654	0.203	712	0.142
598	0.326	656	0.200	714	0.142
600	0.320	658	0.198	716	0.140
602	0.314	660	0.198	718	0.140
604	0.308	662	0.195	720	0.139
606	0.302	664	0.192	722	0.136
608	0.296	666	0.189	724	0.136
610	0.291	668	0.186	726	0.135
612	0.286	670	0.184	728	0.133
614	0.280	672	0.181	730	0.132
616	0.276	674	0.180	732	0.132
618	0.271	676	0.177	734	0.130
620	0.266	678	0.174	736	0.127
622	0.262	680	0.173	738	0.127
624	0.259	682	0.171	740	0.126
626	0.253	684	0.170	742	0.124
628	0.250	686	0.166	744	0.125
630	0.247	688	0.165	746	0.122
632	0.243	690	0.162	748	0.123
634	0.240	692	0.160	750	0.121
636	0.234	694	0.159		
638	0.231	696	0.155		

APPENDIX 3

UV-VIS ABSORPTION SPECTRUM OF SILVER NANOPARTICLES PREPARED USING BIOLOGICAL SYNTHESIS

wavelength	Absorbance
360	1.997
362	1.021
364	1.040
366	1.068
368	1.083
370	1.109
372	1.116
374	1.130
376	1.143
378	1.171
380	1.183
382	1.193
384	1.210
386	1.216
388	1.234
390	1.258
392	1.271
394	1.283
396	1.294
398	1.311
400	1.322
402	1.333
404	1.344
406	1.359
408	1.383
410	1.397
412	1.414
414	1.426
416	1.433
418	1.428
420	1.432
422	1.423
424	1.415
426	1.401
428	1.391
430	1.382
432	1.381
434	1.376
436	1.378
438	1.375
440	1.380
442	1.379
444	1.381
446	1.380
448	1.377
450	1.375
452	1.372
454	1.365
456	1.357
458	1.351
460	1.345
462	1.337
464	1.329
466	1.322
468	1.314
470	1.305
472	1.295
474	1.289
476	1.279
478	1.270
480	1.260
482	1.250
484	1.235
486	1.224
488	1.211
490	1.198
492	1.184
494	1.169
496	1.154
498	1.138
500	1.122
502	1.108
504	1.090
506	1.076
508	1.059
510	1.044
512	1.029
514	1.013
516	0.999
518	0.983
520	0.969
522	0.953
524	0.940
526	0.926
528	0.912
530	0.899
532	0.886
534	0.875
536	0.861
538	0.850
540	0.838
542	0.826
544	0.814
546	0.801
548	0.790
550	0.778
552	0.767
554	0.755
556	0.743
558	0.732
560	0.722
562	0.712
564	0.702
566	0.694
568	0.685
570	0.676
572	0.670
574	0.663
576	0.656
578	0.648
580	0.642
582	0.636
584	0.630
586	0.622
588	0.615
590	0.609
592	0.602
594	0.597
596	0.590
598	0.584
600	0.578
602	0.573
604	0.568
606	0.562
608	0.559

610	0.554	676	0.457	742	0.370
612	0.551	678	0.455	744	0.368
614	0.547	680	0.454	746	0.365
616	0.543	682	0.451	748	0.363
618	0.539	684	0.450	750	0.359
620	0.537	686	0.445	752	0.357
622	0.534	688	0.444	754	0.355
624	0.530	690	0.441	756	0.352
626	0.526	692	0.437	758	0.350
628	0.523	694	0.433	760	0.347
630	0.522	696	0.429	762	0.346
632	0.518	698	0.428	764	0.344
634	0.516	700	0.426	766	0.340
636	0.511	702	0.423	768	0.338
638	0.512	704	0.421	770	0.336
640	0.507	706	0.418	772	0.335
642	0.504	708	0.416	774	0.332
644	0.502	710	0.414	776	0.328
646	0.498	712	0.411	778	0.327
648	0.495	714	0.408	780	0.326
650	0.492	716	0.405	782	0.323
652	0.489	718	0.403	784	0.320
654	0.488	720	0.399	786	0.318
656	0.483	722	0.397	788	0.315
658	0.482	724	0.394	790	0.314
660	0.479	726	0.392	792	0.312
662	0.476	728	0.389	794	0.310
664	0.475	730	0.386	796	0.307
666	0.472	732	0.384	798	0.306
668	0.469	734	0.380	800	0.303
670	0.467	736	0.379		
672	0.464	738	0.375		
674	0.460	740	0.373